A BUSINESS APPROACH TO CORN FARMING

Complete Entrepreneurial Step By Step Guide To Corn Garden From Scratch

ZHURI HART

DISCLAIMER

This book is intended to provide general information and insights on adopting a business approach to farming. The content within is based on the author's knowledge and experiences up to the date of publication. It is essential to recognize that the field of agriculture is dynamic, influenced by various factors such as market conditions, climate, and regulatory changes.

Readers are advised to conduct thorough research, seek professional advice, and consider their unique circumstances before implementing any strategies or practices discussed in this book. The author and publisher disclaim any responsibility for the accuracy, completeness, or suitability of the information provided. The book is not a substitute for professional advice, and the author and publisher shall not be liable for any damages or losses arising from the use or reliance on the information presented herein.

Individual results may vary, and success in farming enterprises is contingent upon numerous variables. The author encourages readers to consult with relevant experts, agricultural extension services, and legal or financial professionals to tailor strategies to their specific needs and local conditions.

This book is not intended to be a comprehensive guide to all aspects of farming, and readers should exercise their judgment and discretion in applying the principles discussed. The author and publisher do not endorse any specific products, services, or companies mentioned in this book unless explicitly stated.

By reading this book, the reader acknowledges and accepts the inherent uncertainties in agricultural endeavors and agrees to use the information at their own risk.

TABLE OF CONTENTS

ABOUT THE BOOK

"A Business Approach to Corn Farming" is a thorough manual created to give prospective and current corn farmers the skills and know-how required to turn their agricultural pursuits into profitable, long-lasting enterprises. This book's significance stems from its comprehensive approach, which covers all the important facets of the corn farming sector—from historical viewpoints to the dynamics of the market today—as well as planning, crucial farming techniques, financial management, marketing tactics, and embracing sustainability and innovation.

By outlining the book's history, objectives, target readership, and boundaries, the introduction section lays a strong foundation. This lays the groundwork for readers to comprehend the content's relevance and how, in the context of corn farming as a company, it meets their unique requirements and interests.

The book explores the vital task of comprehending the grain sector. Readers obtain important insights into the role of maize in agriculture and the larger economy by reading this historical overview, which is followed by an analysis of contemporary developments and market dynamics. The local and international market analysis also contributes to the development of a thorough viewpoint for well-informed decision-making.

Readers are guided through the complexities of organizing a profitable maize farming enterprise in. It lays the foundation for a well-thought-out business plan by highlighting the significance of establishing specific goals, carrying out feasibility studies, choosing appropriate locations, and taking legal and regulatory issues into account. Creating a company plan ensures a hands-on approach by bringing pragmatism to the theoretical components.

The ensuing chapters provide a thorough examination of marketing tactics, sustainable and creative agricultural methods, financial management strategies, and key corn farming practices. Every chapter has been painstakingly designed to offer practical advice, whether it is about selecting the best crop varieties, putting sustainable agricultural methods into reality, or overcoming the particular financial difficulties associated with corn farming.

The book offers workable answers while acknowledging the difficulties faced by the maize farming sector. The information is designed to enable farmers to effectively manage obstacles, from risk mitigation techniques to market change adaptation. Furthermore, the discourse around governmental policies and support mechanisms provides a sophisticated comprehension of the exogenous variables influencing the sector.

"A Business Approach to Corn Farming" is essentially a priceless manual for anyone looking to start, grow, or improve a corn farming business. The book gives readers the tools and techniques they need to not only survive but also prosper in the competitive and dynamic maize-growing sector by fusing theoretical knowledge with real-world applications.

CHAPTER ONE

CORN FARMING INTRODUCTION

CONTEXT

The history of corn growing is extensive, dating back hundreds of years to the earliest American civilizations. Maize, the scientific name for corn, was first cultivated and domesticated by indigenous communities, mostly in Mesoamerica. Corn developed as a staple crop over time, used for trade and cultural purposes in addition to local subsistence. Corn varieties and production methods have changed over centuries as a result of climate shifts and advancements in agricultural science.

A pivotal moment in the history of agriculture occurred when explorers such as Christopher Columbus brought maize to the European peninsula. The global acceptance of corn can be attributed to its adaptability and versatility. Corn is now a staple of the world's food supply, used as a main component in a wide range of culinary products, animal feed, and industrial uses.

SYNOPSIS OF CORN FARMING AS AN ENTERPRISE

Corn production has developed into a vibrant, diverse industry that is essential to the agricultural industry. There are several steps involved in growing corn, from choosing the right location and preparing the soil to planting, tending to, and harvesting. To optimize harvests and reduce their impact on the environment, farmers use both traditional and modern farming techniques, including cutting-edge technologies and sustainable farming approaches.

Corn farming is a business enterprise that involves more than just the actual cultivation. The economic sustainability of corn farming operations is heavily influenced by pricing patterns, market dynamics, and worldwide demand. To maintain their competitiveness and guarantee a successful crop, farmers must manage these variables in addition to taking into account developments in agronomic techniques, pest control, and seed technology.

Corn farming is a company that involves more than just the agricultural sector; it also involves marketing, distribution, and supply chain management. Corn is a crop that can be used in a variety of industries.

Its products are used in animal feed, food manufacturing, biofuel production, and other industrial processes. Successful corn farmers are skilled at negotiating the market's intricacies, comprehending consumer tastes, and adjusting to the changing needs of a globalized economy.

Corn cultivation involves not just economic factors but also resource management and environmental sustainability. Sustainable farming practices are being adopted by farmers on a larger scale to minimize their environmental impact, preserve soil health, and maximize water use. Corn farming firms are more efficient and sustainable when technology, precision agriculture, and data-driven decision-making are combined.

The history of corn growing is rich, and the way it developed into a contemporary industry illustrates how resilient and flexible agricultural methods can be. A thorough examination of the historical background of corn farming, the nuances of cultivation, and the larger economic and environmental factors that influence the sector are all necessary to comprehend the business's complexity.

CHAPTER TWO

COMPREHENDING THE CORN INDUSTRY

AN OVERVIEW OF CORN FARMING HISTORY

The history of corn also referred to as maize, is intricately linked to the advancement of agriculture. Corn is an American crop that has long been a staple for Native American cultures, even before European settlers arrived. After the Columbian Exchange, its cultivation expanded around the world and became an essential component of diets everywhere. Corn was first cultivated in the United States by Native American cultures, who then adopted and spread the practice across Europe.

Maize cultivation was transformed in the early 20th century with the introduction of hybrid maize cultivars. Enhancing crop yields and disease resistance was made possible in large part by scientific discoveries, notably the efforts of scientists like George Washington Carver. The sector grew as a result of the advancement of machinery, such as the corn combine harvester, which significantly increased the efficiency of harvesting maize.

MARKET DYNAMICS AND CURRENT TRENDS

The maize sector is characterized by several noteworthy trends and dynamics in the modern environment. Genetically modified organisms (GMOs) and precision agriculture are two examples of technological innovations that have greatly enhanced maize yields and resistance to pests and illnesses. Environmental concerns are driving a growing trend in farming methods adoption that affects both large agribusinesses and small-scale farmers.

Because of its many uses, corn has seen an increase in demand worldwide. Apart from its customary application as a staple grain, maize plays a critical role in the manufacturing of animal feed, biofuels, and other industrial goods. Governments all around the world are putting more emphasis on renewable energy sources, and this has made the ethanol sector in particular a significant driver of maize demand.

CORN'S SIGNIFICANCE FOR AGRICULTURE AND THE ECONOMY

The importance of corn to agriculture and the wider economy is complex. Corn is a key cereal crop that provides both people and animals with their main source of sustenance. Its adaptability covers a wide range of industries, such as manufacturing, biofuel generation, and food processing. Corn is a major component of animal feed used by the livestock industry, which emphasizes its importance in sustaining the meat and dairy sectors.

The maize industry makes a significant economic contribution to the GDP of corn-producing regions. Corn cultivation affects both large agricultural operations and small-scale farmers by creating job possibilities in rural areas. Furthermore, the export of maize and items generated from it is essential to global trade because it strengthens the bonds between countries that produce and consume corn.

INTERNATIONAL AND REGIONAL MARKET ANALYSIS

Understanding both local and global dynamics in detail is necessary for analyzing the maize market. Among the world's leading producers of corn are the US, Brazil, China, and Argentina, all of which make substantial contributions to the world supply. Weather patterns, trade laws, and technical developments all have a direct impact on world corn prices and supply networks.

Differentiated market circumstances are created locally by regional differences in soil types, climate, and farming practices. Local maize markets are significantly shaped by government policies, subsidies, and regulations. In certain areas, consumer preferences and market dynamics are being impacted by the move toward sustainable agriculture and the promotion of organic corn-growing practices.

CHAPTER THREE

ORGANIZING YOUR CORN FARMING ENTERPRISE

CLEARLY DEFINING YOUR OBJECTIVES AND GOALS

A critical first step in organizing your corn-growing business is establishing specific goals and objectives. These objectives form the cornerstone of your whole business, directing decision-making procedures and offering a path to prosperity. Having well-defined goals aids in concentrating on important elements like production targets, market share, and financial benchmarks. Establishing a family farm that is sustainable or becoming a large supplier in your area are two different goals that you may pursue, but both require well-defined targets to plan well.

PUTTING TOGETHER A FEASIBILITY STUDY

An additional crucial step in getting ready for your corn farming endeavor is to conduct a feasibility analysis. In-depth analyses of several variables, such as resource availability, market demand, and potential difficulties, are part of this study. By evaluating your business idea's viability, you may see potential hazards and possibilities and make sure that your efforts are supported by a realistic grasp of the market and industry dynamics. You can improve your company model and make well-informed choices regarding technology, investments, and operational tactics with the aid of a thorough feasibility study.

SELECTING THE IDEAL SITE

Selecting the ideal site for your corn farm is a calculated move that will have a big effect on how successful your company is. It is important to give serious thought to elements including soil quality, climate, and market accessibility. Infrastructure related to transportation and access to water resources are other important factors. By carefully weighing these variables, you can be sure that the place you've selected will meet your production needs and offer a favorable atmosphere for your corn farming activities.

REGULATORY AND LEGAL ASPECTS

Establishing and running a corn farm involves many legal and regulatory considerations. Learn about the rules that apply to land use, business licensing, agriculture, and environmental protection at the municipal, state, and federal levels. Maintaining your farming business's viability and avoiding legal problems depend on your compliance with these regulations. You can manage the complexity of regulatory obligations by seeking advice from agricultural extension services or legal specialists.

CREATING A BUSINESS STRATEGY

A well-crafted business strategy is the foundation of any successful corn farming operation. A well-written business plan includes operational specifics, financial estimates, and an overview of your overall approach.

It is a useful tool for getting funding from banks or investors and acts as a roadmap for your company. Important topics such as production techniques, marketing plans, risk assessment, and financial analysis should all be included in your company plan. You may adjust your business plan to changing market conditions and drive your corn farming firm toward long-term success by reviewing and revising it regularly.

CHAPTER FOUR

CRUCIAL TECHNIQUES FOR GROWING CORN

CROP VARIETY AND SELECTION

The proper selection of crop variety is one of the key elements of successful corn cultivation. A few of the traits that differ amongst corn types are maturity period, disease and insect resistance, and soil and climate adaptation.

When selecting the best variety of corn, farmers need to take into account things like the climate in the area, the type of soil, and how the produced corn will be used. This guarantees that the crop of choice will flourish in the specified conditions and fulfill the required production and quality requirements. Optimizing corn farming output can also be achieved by keeping up with developments in hybrid varieties and corn genetics.

FERTILIZATION AND SOIL PREPARATION

Creating the ideal environment for corn development requires effective soil preparation. To evaluate soil composition and nutrient levels and make educated judgments about fertilization, farmers must do soil testing. The fertility and structure of soil are enhanced by the addition of organic matter, such as manure or compost. A key part of soil preparation is leveling the ground, ensuring proper drainage, and plowing the ground correctly. Applying fertilizers such as potassium, phosphorus, and nitrogen sparingly depends on the needs of the corn crop and the findings of soil tests. This meticulous method of fertilizing and preparing the soil greatly enhances the general well-being and yield of the maize plants.

PLANTING METHODS AND SCHEDULE

One important component affecting corn output is planting precision. When planting corn, farmers need to take into account variables including plant population, seed depth, and row spacing. Because maize is temperature- and light-sensitive, planting time is also very important. The development of the crop and germination might be impacted by planting too soon or too late. Furthermore, using contemporary planting technologies—like precision planting tools—can improve the precision and efficiency of seed placement. Crop rotation techniques are another option for farmers looking to disrupt pest and disease cycles and enhance soil quality for upcoming corn harvests.

SYSTEMS OF IRRIGATION

For corn to reach all of its growth stages there must be sufficient and steady moisture. Installing effective irrigation systems is crucial, particularly in areas with erratic rainfall patterns. A farmer's options for irrigation can vary based on land terrain and water availability and can include furrow irrigation, center pivot systems, and drip irrigation. To prevent under or overwatering, proper irrigation scheduling is essential, taking into account variables like crop growth stages and soil moisture levels. The effectiveness of irrigation systems in corn farming can be further increased by putting water-saving techniques like mulching into practice.

MANAGEMENT OF PESTS AND DISEASES

Numerous pests and diseases can have a substantial impact on the yield and quality of corn crops. To effectively manage pests, Integrated Pest Management (IPM) strategies combine chemical, biological, and cultural control methods.

Using resistant cultivars, rotating crops, and ensuring appropriate plant spacing are cultural methods that aid in pest management. In biological control, natural predators of the pests are introduced; in the case that chemical control is required, this should be done carefully and with consideration for the effects on the ecosystem. Regular field scouting is essential for prompt intervention and efficient management to spot pests and illnesses.

HARVESTING AND HANDLING AFTER HARVEST

Corn harvesting must be done at the right time to guarantee optimal quality and yield. The quality of the kernels and total yield might be affected by harvesting too soon or too late. Harvesting can be done quickly and efficiently thanks to modern harvesting tools like combines. Harvested corn must be dried, cleaned, and stored as part of post-harvest management. For mold to not grow and to lower the moisture content, proper drying is necessary. By cleaning the maize, extraneous elements are removed and quality standards are met. Sufficient storage spaces, like bins or silos, are required to shield the corn from pests and the elements. Maintaining the quality of the produced corn and increasing its market value requires the use of effective post-harvest measures.

CHAPTER FIVE

MANAGING MONEY IN CORN FARMING

COST ANALYSIS AND BUDGETING

Cost analysis and budgeting are essential components of corn-growing businesses' financial management. The methodical planning and distribution of financial resources to accomplish particular objectives is a component of budgeting. Making a thorough budget for corn growing enables farmers to project the expenditures of several cultivation-related expenses, such as labor, machinery, irrigation, fertilizers, pesticides, and seed. Farmers may better manage their finances, find areas where costs can be cut, and make decisions that maximize resource use by carefully laying out these expenses.

By offering a thorough breakdown of costs, cost analysis helps farmers better understand the cost structure of their operations and is a useful tool in conjunction with budgeting. T his research makes it possible to pinpoint cost-effective procedures, cost drivers, and locations where efficiency gains can be made. Corn producers may improve their budgeting procedures, boost operational effectiveness, and eventually increase profitability by routinely monitoring and assessing expenditures.

FORECASTING REVENUE

Another crucial component of corn farming's financial management is revenue forecasting. Farmers may anticipate income levels and make well-informed decisions about production, pricing, and marketing tactics with the aid of precise revenue forecasts. Revenue is influenced by several factors, including commodity pricing, market circumstances, and crop yields. Farmers must constantly evaluate these factors and modify their estimates as necessary. Corn growers may improve the precision of their income projections and make well-timed strategic decisions by leveraging market trends, historical data, and industry insights.

RISK MANAGEMENT IN FINANCE

In the volatile agricultural business, where outside variables like weather, market swings, and policy changes can have a big impact on results, financial risk management is essential. To reduce possible financial losses, farmers can use a variety of risk management techniques, such as insurance, futures contracts, and diversification. Corn producers can create proactive plans to protect their financial interests and guarantee the sustainability of their operations by identifying and evaluating potential hazards.

GETTING GRANTS AND FUNDING

Obtaining funding and grants is an important factor for corn farmers who want to grow their businesses, make investments in new technology, or deal with unforeseen difficulties. To find more money, farmers might look at several options, such as government grants, agricultural loans, and subsidies. Gaining access to these financial resources requires building solid relationships with banking institutions, governmental bodies, and groups that promote agriculture. Corn farmers can increase productivity by investing in modern equipment, using sustainable methods, and obtaining grants and funding if they are successful in doing so.

MAINTAINING DOCUMENTS AND FINANCIAL REPORTING

The cornerstones of efficient financial management in corn farming are record-keeping and financial reporting. Farmers can keep track of their revenue, expenses, and other financial transactions by keeping precise and current records. Budgeting, cost analysis, and regulatory compliance all benefit greatly from this data. Furthermore, financial reporting offers clear insights into the farm's financial health to all relevant parties, such as lenders, investors, and government authorities.

CHAPTER SIX

CORN PRODUCT MARKETING STRATEGIES

MARKET DEMAND UNDERSTANDING

A thorough grasp of market expectations is essential for corn product marketing to be effective. To determine consumer preferences, trends, and prospective market gaps, extensive market research must be conducted. Understanding the various uses for maize products—like cornmeal, corn oil, and corn-based snacks—allows companies to customize their marketing approaches to target certain customer demands.

A successful marketing strategy depends on matching offerings with market demands, whether the target audience is health-conscious consumers looking for gluten-free options or those interested in locally sourced and sustainable products.

BUILDING YOUR CORN PRODUCTS' BRAND

Developing a good brand is essential to corn products' commercial success. The brand ought to represent the principles, excellence, and distinctiveness of the corn goods. This entails crafting an engaging brand narrative that appeals to customers and sets the products apart from rival offerings. The entire image of a brand is greatly influenced by factors including product placement, brand message, and packaging design. The consumer image of a brand can be positively impacted by highlighting elements like the company's dedication to sustainability, the nutritional advantages of maize, or the use of non-GMO corn.

PRICING STRATEGIES

The competitiveness of corn goods in the market depends on choosing the appropriate pricing strategy. It is important to carefully analyze factors including production costs, market demand, and rival pricing. Whether promoting the products as premium, midrange, or affordable, price positioning can be done deliberately. Customers can also be drawn in and kept by using dynamic pricing tactics based on seasonality or promotional times. Achieving a balance between providing value for money and sustaining profitability is crucial to make sure that the pricing plan is in line with the target market's perception of the worth of the corn goods.

DISTRIBUTION ROUTES

One of the most important aspects of marketing corn goods is choosing efficient distribution routes. This entails assessing a range of channels, including direct-to-consumer models, internet platforms, specialized shops, and supermarkets. Determining the best distribution channels requires an understanding of the buying tastes and habits of the target audience. Forming alliances with distributors, merchants, and internet sellers can increase corn products' accessibility and exposure. The timely and effective delivery of the items to the appropriate market groups is guaranteed by a carefully thought-out distribution strategy.

DEVELOPING CONNECTIONS WITH PURCHASERS

Marketing corn products must establish trusting relationships with consumers if it is to be successful in the long run. This calls for clear communication, comprehension of the demands of the consumer, and top-notch customer support. Whether they are companies or individual customers, having open lines of communication with them facilitates feedback collection and timely problem-solving. Providing incentives to customers, such as exclusive promotions, loyalty programs, or bespoke product offerings, can encourage customer loyalty. Creating and sustaining good relationships promotes favorable word-of-mouth advertising, which is crucial in the cutthroat food industry. It also encourages return business.

CHAPTER SEVEN

INNOVATIVE AND SUSTAINABLE PRACTICES

PRACTICES OF SUSTAINABLE AGRICULTURE

The use of Sustainable Agriculture Practices is essential in mitigating the issues raised by traditional farming techniques. Sustainable agriculture prioritizes the long-term health of the environment, social justice, and economic sustainability to maximize resource usage while reducing adverse effects. Crop rotation, agroforestry, and organic farming are examples of practices that increase soil fertility, lessen dependency on artificial inputs, and foster biodiversity. These methods improve agricultural systems' resistance to climate change while simultaneously protecting ecosystems.

USING TECHNOLOGY IN CORN PRODUCTION

Using Technology in Corn Farming has emerged as a major force for innovation and sustainability in the field of modern agriculture. For example, precision farming makes use of cutting-edge technologies like GPS, sensors, and data analytics to maximize the use of inputs like water, fertilizer, and pesticides. Minimizing resource usage decreases environmental effects in addition to optimizing crop production. Furthermore, the use of genetically modified (GM) crops has the potential to increase agricultural output overall, decrease the need for chemical inputs, and improve crop resilience.

ECOLOGY PRUDENCE

An all-encompassing strategy for properly managing ecosystems and natural resources is environmental stewardship. It comprises realizing how closely human activity is linked to the environment and taking action to reduce adverse effects. Environmental stewardship in agriculture refers to applying techniques that maintain biodiversity, cut greenhouse gas emissions, and maintain water quality. This stewardship includes conservation tillage, cover crops, and sustainable water management, all of which promote a peaceful coexistence between agricultural production and the surrounding ecosystems.

ACCREDITATIONS AND STANDARDS OF QUALITY

Ensuring that innovative and sustainable practices are followed across the agricultural value chain is made possible in large part by certifications and quality standards. A reputable certification, like Rainforest Alliance or USDA Organic, gives customers peace of mind that the goods they buy adhere to strict social and environmental standards. Respecting these guidelines improves fair labor practices, encourages farmers to adopt sustainable techniques, and builds the agricultural industry's credibility overall. These certifications help the environment and foster customer confidence in the sustainability of agricultural products.

The agricultural industry becomes more resilient, efficient, and responsible when Sustainable Agricultural Practices, Technology Integration in Corn Farming, Environmental Stewardship, and Compliance with Certifications and Quality Standards are all integrated. Adopting these ideas is essential for the long-term sustainability of our food systems, as the globe confronts more and more issues about environmental degradation and food security. The secret to turning agriculture into a global force for good is to combine conventional wisdom with state-of-the-art technology.

CHAPTER EIGHT

PROBLEMS AND REMEDIES IN CORN PRODUCTION

TYPICAL INDUSTRY CHALLENGES

Like any other agricultural sector, corn growing has several difficulties that need creative solutions to maintain productivity and sustainability. Climate fluctuation is a common difficulty faced by the maize farming sector. Crop production can be greatly impacted by erratic weather patterns, such as droughts, floods, and extremely high or low temperatures. It is imperative to develop resilient and climate-smart farming systems since farmers frequently find it difficult to adjust their practices to these shifting conditions.

STRATEGIES FOR RISK MITIGATION

The ubiquity of illnesses and pests is another difficulty. If not adequately managed, corn crops are vulnerable to a wide range of insects and viruses that can destroy harvests. To meet this problem, integrated pest management solutions that combine biological control techniques with sparing chemical applications are crucial. Furthermore, continuing studies on resistant maize types can provide a more environmentally friendly method of controlling pests and diseases.

In addition, farmers deal with problems related to soil health, such as soil erosion and nutrient depletion. Ongoing monoculture techniques may result in nutrient imbalances in the soil, which may affect the land's long-term fertility. Crop rotation, cover crops, and sustainable soil management techniques can all be used to mitigate these issues and enhance the general health of the soil.

A vital component of maize cultivation is risk minimization because agricultural markets are unstable. Farmers may face financial risks due to fluctuating commodity prices, input costs, and outside variables like geopolitical events.

Farmers frequently use risk management techniques, such as hedging and crop insurance, to offset these risks. With a certain amount of financial security offered by these instruments, farmers can handle market swings with more assurance.

CHANGING WITH THE MARKET

The corn farming industry is always faced with the problem of adapting to changes in the market. Population increase, dietary changes, and the growing use of maize in the manufacture of biofuels are some of the variables that can impact the demand for corn and its derivatives globally.

To satisfy changing consumer expectations, farmers need to be aware of market trends and adaptable in how they modify their production methods. A farmer can become more adept at responding to market changes by diversifying their product line, pursuing niche markets, and adopting sustainable agricultural methods.

GOVERNMENTAL GUIDELINES AND ASSISTANCE

The funding and policies provided by the government greatly influence how corn growing is practiced. Trade agreements, environmental restrictions, and subsidies all have a big influence on how profitable and sustainable corn farming may be. Farmers can be empowered to implement creative and sustainable techniques by having access to financial incentives, research money, and extension services. Therefore, cultivating a robust and prosperous maize farming sector requires a supporting policy environment.

The difficulties encountered by maize growers are diverse and include market dynamics, regulatory effects, soil health issues, pest and disease pressure, and climate variability. It takes an all-encompassing and integrated strategy that incorporates strategic risk management, sustainable farming methods, and technology developments to address these issues.

www.ingramcontent.com/pod-product-compliance
Lightning Source LLC
Chambersburg PA
CBHW070839290526
45795CB00002B/921